Chicken Hot Topics

Editor: Dr. Katherine Mansfield-Houle
Cover Image Courtesy of: Anja Osenberg

CHICKEN HOT TOPICS
CONTROVERSIAL HUSBANDRY
PRACTICES

JESSICA E. LANE

First Edition, 2014
Published in the United States of America
ISBN: 978-1502538123

This book is dedicated
to all of the special friends
who encouraged me
to take a leap of faith.

I want to send out a special thank you to the individuals who made writing this book possible. My husband has been so supportive during this process and has encouraged me to press on any time I have felt overwhelmed with the task.

My dear friends Kathie Lapcevic, Tessa Zundel, Angi Schneider, and Chris Dalziel, I am so glad to have such wonderful ladies in my life. You have been so amazingly supportive of not only this book, but of other aspects of my life as well.

My dearest Katie, you did more than edit a book. You gave my words the direction and clarity they needed to reach people. You helped me reach my goal.

To all of you, I dedicate this book.

Table of Contents

INTRODUCTION

When I first got chickens, I searched high and low to find out as much about my new feathered creatures as I could. My parents raised chickens for a short time when I was a child, but my only memories of the experience involved the harsh lesson that you don't stick your fingers in the pen or the chickens will peck you. As an adult, I was entering uncharted territory and I wanted to do the best I possibly could.

The problem I had when I first began my poultry keeping experience was my finding all of the conflicting information online. One site would say *this* is the best choice for your flock, but this other site would say *that* is the better choice. How was I to know what to do? I looked for research that had been done on chicken husbandry. I looked for studies that confirmed or disputed what I had read in the literature I found. I was left with too many information gaps and unanswered questions.

I decided to sit down and write <u>Chicken Hot Topics</u> with the desire to write a fact-based book to assist you in choosing what is best for your flock. I wanted to use evidence-based data to dispel myths that float around the internet. I wanted poultry keepers to be equipped to make sound, logical decisions for the well-being of their flock without feeling influenced by my personal feelings.

CHICKEN HOT TOPICS

The topics in this book are wide-ranging and include the most controversial concepts in poultry husbandry. In addition to the facts, you will find emotion on the pages of this book. As human beings, how we feel is one of the tools we use to make our life decisions. I trust you are reading <u>Chicken Hot Topics: Controversial Husbandry Practices</u> with both intellectual curiosity and because you care about the welfare of your chickens. I hope this book arms you with the knowledge to make the best decisions for you and your flock.

Husbandry *[n] The raising of domestic animals.*

SPACE REQUIREMENTS

There is a huge range of possibilities when it comes to the space required for raising chickens in a healthy environment. Cramped spaces can lead to health and behavioral issues in your flock. With a recommended range from three square feet per bird to more than sixteen, how are we suppose to know what size is ideal? Do you need one nest box for every three hens or for every five? Much like choosing the appropriate bedding, space requirements are contingent upon your maintenance routine and climate, as well as your flock.

If you have extreme temperatures on any end of the spectrum (hot or cold; moist or dry), you will need more indoor space to accommodate birds that prefer to stay indoors and away from the elements. If your flock spends the majority of their time free ranging, you can squeak by with less indoor space. If you meticulously clean your coop, you can maintain a smaller space without fear of ammonia buildup or unsanitary conditions.

General Guidelines

Storey's Guide to Raising Chickens (2010) puts it well when they state, "*Because chickens spend most of their active time outside of the coop, generally 2-3 square feet per chicken is sufficient space. If you plan on keeping them cooped up full-time then 8-10 square feet per chicken would do, counting the outside run. If you have 25 hens you don't need to purchase 25 individual nest boxes. In fact, one six-hole nest box would probably be sufficient for 25 laying hens."* These figures are general guidelines. As stated previously, every poultry keeper has a different flock and unique environment to consider.

Illustration 1: Often times multiple hens will crowd into one nest box, even if other boxes are available.

Watch your birds. Be aware of behavioral issues such as lethargy, reduction in egg production, and signs of respiratory distress. All of these behaviors may be indications that your space is too small for the number of chickens you are housing. Slowly add birds until you reach the magic number for your coop. It is better to have too few than too many.

Build Smart

When you start your poultry keeping experience, I encourage

you to build and design with as much space as you are able to provide. Not only can you easily downsize your space if you desire, but you will be prepared in the event Chicken Math gets the best of you. I began my poultry keeping experience with a 3' x 4' chicken coop which well-housed four hens named after The Golden Girls. My current setup consists of an 8' x 8' remodeled shed. It contains five nest boxes and comfortably houses fifteen chickens and a trio of large breed ducks, even through our long, cold Maine winters.

FUN FARMER FACT

Chicken Math is a calculated method to rationalize adding to your flock. An example would be purchasing a coop that houses three birds, but ordering five to meet shipping minimums. Doubling the size of the coop so it now holds six birds, then feeling the desire to order more birds because you have a space that needs to be filled. But wait! There's a minimum of five, so you need a bigger coop. Chicken Math is like an addiction and it can become a never ending cycle. Take it from the woman who only wanted four production breeds and currently has a flock of eighteen birds.

HUSBANDRY HINT

Choose individual nest boxes that can be installed or removed as needed, such as these:

There are also great DIY plans available online.

ALTERNATIVE BEDDING

Bedding is an essential element in poultry keeping. Poultry keepers' loyalties are often clearly defined and they share with passion what works best for them and their flock. Many share their "horror stories" from bedding options that failed to meet their expectations to dissuade others from wasting their time. I've seen warnings on poultry forums like "*Don't use straw! I did and my whole flock was infected with scaly mites. It took months to get rid of the mites.*" The warnings seem well-intentioned, but need to be read with a critical eye. No one wants mites on their chickens, but did the keeper maintain the bedding properly? Did they introduce a new bird who may have brought mites with it?

Everyone is going to have different experiences because our situations are different. I encourage you to research and look for what works best in your climate. Moisture may be an issue for those that live in a damp, rainy climate. Those that live in a cold climate may need a bedding with insulating properties. Ask local poultry keepers what they use and how they maintain it.

An alternative bedding option is gaining popularity and may be one of the most controversial husbandry practice out there: Sand. Poultry keepers have debated traditional straw, hay, and shavings for decades. Sand is deemed new and modern. One thing all traditional keepers seem to agree on is that modern has no place in traditional farming. Many bloggers are quick to advocate for the use of sand, but they fail to provide the research behind the safety of this bedding choice. There are bloggers who denigrate the use of sand, but they fail to provide research that shows this bedding choice is unsafe. Online poultry forums seem to be evenly divided on the topic of sand. Members are strong advocates for using it or they are adamantly against it.

Wood Shaving Litter

Wood shavings have always been the standard when it comes to poultry house bedding. Cost and availability have been the deciding factors for most poultry keepers, but with wood shavings now being used to make fiberboard, paper, cardboard and more, the demand for wood shavings is increasing the price. A number of universities, including the University of Kentucky and the University of Alabama, have been studying the use of sand as an alternative litter. Most of this research has been conducted with broiler flocks, but the results of using sand as a substrate applies to layer flocks as well.

Sand as a Substrate

Research indicates that broilers raised on sand grew and developed as well if not better than those raised on pine shavings and other organic materials.[1] The Alabama Agricultural Experiment Station researchers state that the primary advantages of sand are "*less caking, longer use time, less organic decomposition and build-up, and unique horticultural characteristics as a soil amendment.*"[2] Sand's weight makes it less likely to end up scratched into open water containers. Were sand to

8

get into an open water container, the weight would sink it to the bottom, preventing water spoilage. Sand does not require a full clean-out every two weeks to six months like many organic options. With proper maintenance, sand can provide a safe and healthy bedding for a year or more. Sand is a great option if you garden on your property. You can create a large sifter (similar to a cat-litter scoop) and separate excretions from the bedding, allowing for direct application to the compost bin without other elements that may require breaking down. Sand tends to stay 2° cooler in the summer and 2° warmer in the winter when used in an enclosed building, such as a chicken coop. That thermal difference makes sand a great choice for the hottest climate to the coldest climate and everywhere in between.

In addition to the advantages above, additional studies have shown that foot pad quality is improved with the use of sand and there are fewer respiratory issues in broilers due to lower dust levels.[3] I use sand and have had no incidents of bumble foot – a bacterial infection with inflammation of the feet of birds – since switching to sand, even with my plus-sized ladies.

According to the Poultry Science Association, Inc., from a bacteriological standpoint, sand is equal to or slightly superior to organic materials when used as bedding. E. coli and plate counts were significantly lower in sand. Although moisture and ammonia levels were similar to those in pine shavings, there were significantly lower numbers of bacteria found in the sand bedding.[4]

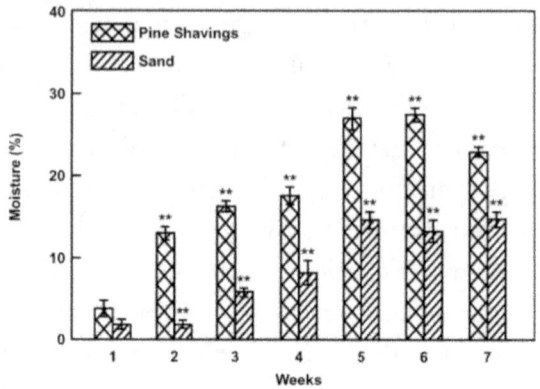

Table 1: The percentage of moisture present in pine shavings and sand according to the study conducted by the Poultry Science Association, Inc.

Organic materials, such as wood shavings, straw and hay, tend to contain aspergillus organisms. Aspergillosis, which spawns from aspergillus spores, is a fungal infection that affects most frequently the respiratory system of chickens. The Poultry Site reports aspergillosis morbidity in adult birds is relatively low (12%), but may cause chronic lesions on the lungs. In young birds, morbidity levels range from 5-50%.[5]

Rice Hull Litter

The bedding you choose, of course, may depend on availability and cost in your area. Rice hulls may be available in some parts of the country and be a cost effective option for poultry keepers without access to sand. Rice hulls are typically free from excessive dust and their size, thermal conductivity, and drying rate make them superior to pine shavings, hay and straw.

HUSBANDRY HINT

Poop boards are a great way to keep your bedding cleaner for longer. Poop boards are simply a board positioned under the roosts and lined with easy-to-clean materials such as newspaper, straw and hay for a quick daily clean up.

CHOOSING THE RIGHT SAND

When choosing sand for your coop, select sand with a variety of grit sizes in it. Bank run sand and construction sand are great options and are available at most quarries or in small quantities at home improvement stores. Play and sandbox sand are buoyant, causing water spoilage and are too fine to act as grit.

1   Jacob, Dr. Jacquie "Sand As an Alternative Litter Material" *Cheeps & Chirps* v 1.4 (2008)

2   "Turning Trash into Treasure: Sand as Bedding Material for Rearing Broilers" *Highlights of Agricultural Research* v 47.1 (2009)

3   Kirkhorn, Dr. Steven R. "Agricultural Respiratory Hazards & Disease" *Partners in Agricultural Health* v 4 (2013)

4   Poultry Science Association, Inc. "Bacterial Levels of Pine Shavings and Sand Used as Poultry Litter" *Oxford Journals* (2014)

5   The Poultry Site Guide to Poultry Diseases "Aspergillosis" *http://www.thepoultrysite.com/diseaseinfo/7/aspergillosis*

SPECIALTY DIETS

Before the twentieth century, flocks were primarily kept on working farms where their diets consisted of foraged foods and was supplemented with grains, garden waste, kitchen scraps, and calcium supplements like oyster shell and egg shell. As poultry farming became more specialized, flocks grew too large to feed in this mostly scavenging manner. In the last century, researchers have been formulating commercial feeds. The process is relatively new and is often being modified for improvements and cost-efficiency.

Commercial Feed

Corn and soybean are two very prominent and controversial ingredients in commercial feed. Feed contains corn for energy and soybean meal for protein. Both of these ingredients are under serious scrutiny as the battle against genetically modified foods (GMOs) persists. In addition to these controversial ingredients, commercial feeds often contain antibiotics and arsenicals – arsenic-containing compounds – to support health and enhance growth, ccoccidiostats to combat coccidiosis, and sometimes mold inhibitors. The National Sustainable Agricultural Information Service deems these compounds to be safe for poultry health.[1]

CHICKEN HOT TOPICS

If you are uncomfortable with any of these compounds, you can steer clear of them by reading labels and choosing organic commercial feeds.

Organic Feed

At this time, the U.S. Department of Agriculture (USDA) does not sanction the labeling of livestock rations as "organic," however, private and state certifying agents may endorse rations as organic if the feed meets the agency's requirements. To be certified organic by a private or state agent, rations must contain 100% organic ingredients and contain no antibiotics, wormers, growth promotants, or insecticides beyond the ones specified on the program's list of approved products.

Illustration 2: Chickens gathered around the feeder.

Soy-Free Feed

Specialty diets are becoming increasingly popular among

naturalist poultry keepers. A naturalist is one who believes the chicken's diet should not be enhanced or modified artificially. One of these increasingly popular specialty diets is the Soy-Free Diet, which stems from the anti-GMO movement. Proponents of the soy-free diet have concerns about the birds consuming the soy as well as the humans consuming the birds' meat and eggs. Soybean meal accounts for approximately 30% of the commercial feed content and contains not only more than 45% crude protein, but linoleic acid, and essential nutrient in animal diets.[2] Despite moderate research on various grains and oils, no suitable alternative to soy is yet available.

Vegetarian Feed

Another specialty diet that is gaining popularity is the Vegetarian Diet, which relies heavily on soybean meal as the only significant source of protein. Interestingly, neither the Soy-Free nor the Vegetarian Diet suggest options for protein alternatives. Being omnivores, chickens require protein as a significant part of their diets. The vegetarian diet has the additional drawback from the removal of animal products, as they are found lacking amino acids. Some research has been conducted, but at this time no alternative ingredient that provides the essential amino acids but does not throw off nutritional ratios, has been found.

Home-Mixed Rations

Some poultry keepers prefer to mix their own rations to ensure that the feed contains only the ingredients they feel comfortable giving their chickens. Proper balance of protein, grain, minerals and vitamins in homemade diets is essential for the well-being of your birds. Improper balance of nutrients may cause nutritional deficiencies which may lead to diseases. Poultry keepers who are interested in mixing their own feed should refer to The National Research Council's Nutrient Requirements for Poultry to assure a well-balanced diet is being provided.[3]

Feed Themselves Movement

There is a new movement that focuses on chickens feeding themselves with what is available where they graze. Although this movement sounds similar to pasture raising, there is one striking difference. In pasture raising, attention is given to the pasture on which the poultry is being raised. A common practice in true pasture-raised fashion is to use Sudan grass for summer grazing, oats and wheat for winter, and alfalfa as a perennial legume pasture.[4] This "fend for themselves" movement does not seem to take into account proper nutritional balance and may lead to nutritional diseases such as Caged Layer Fatigue (CLF), Rickets, and Fatty Liver Syndrome (FLS).

FUN FARMER FACT
Chickens can't taste sweetness or heat from spicy foods.
They can, however, taste saltiness and tend to avoid it.

HUSBANDRY HINT
Crushed egg shells are a great free calcium source. So long as you refrain from feeding intact shells, there is no reason for your chickens to develop an egg eating habit.

1   Cheeke, Peter R. "Feeds and Feeding" *Applied Animal Nutrition* (1991)

2   "Soy Free Diets For Poultry" *Oregon Small Farms* v 5.3 (2014)

3   The Nutritional Research Council's Nutrient Requirements for Poultry *http://www.nap.edu/*

4   "Feeding Chickens for Best Health and Performance" *The Poultry Site* (2014)

EGG SHELL SUPPLEMENTS

There is a common misconception that feeding chickens egg shells as a source of calcium will increase the likeliness that your chicken will crave this calcium source and begin eating its own eggs. Although this seems like a valid conclusion, it just isn't the case. The reason a chicken begins eating its own eggs is because it has discovered the tasty yoke and white inside. Unless you are feeding your chickens a whole egg, they do not make the connection that these are the same thing.

To assure that your chicken does not make the connection, there are simple steps you can take. Save your egg shells until they have had a chance to dry out. Crush them to assure that they no longer resemble the original egg shape. If you want to dry a large quantity of shells to put in a supplement hopper or dish, you can dry them on a tinfoil lined baking sheet on low in the oven for 30 minutes to an hour. You may then choose to grind them in a blender versus crushing them by hand.

Safety is of course a consideration when you are feeding egg shells to your chickens. It is wise to feed only your own shells because you know the environment your eggs come from. Bacteria may be present on the egg shell not only from the hen itself, but from the environment it's laid in. Your chickens are accustomed to these bacteria, but bacteria from a different flock may cause health issues in your flock.

Store bought eggs should never be fed to chickens as a calcium supplement. U.S. commercial eggs are washed first with an alkaline-based detergent, followed by a sanitizing solution that consists of a chlorine mixture.[1] These compounds are not safe for your chickens to consume.

_____

1   Poultry Science Association, Inc. "Chemical Analysis of Commercial Egg Wash Water" The Journal of Applied Poultry Research v. 6 (2005)

SOLUTIONS TO EGG EATING

Chicken that eat their own eggs use to be given an immediate death sentence, often becoming the next meal on the table. This is still true of farms where egg production is of the utmost importance. For the small-scale poultry keeper, time can be invested in breaking the negative habit of egg eating.

Why it Happens

Egg eating often develops when an egg is accidentally crushed and the chicken(s) discover the goodness contained inside. The trick first to avoid the likelihood of it happening and have a plan in place in case it still occurs. Often a chicken can easily be redirected so long as a habit has not formed. By having a plan in place to circumvent the behavior, there is a good chance you will never need to deal with an egg eater.

Prevention

A soft substrate or bedding is key to preventing eggs breaking. A disposable liner, such as a piece of cardboard, or a washable liner, such as foam shelf liner or commercial nest box liners, act as security in case the substrate has been scratched aside. It assures that the egg still has a padded surface to land on. On top of your liner, you want to have two to three inches of soft bedding.

Cut down the likelihood of multiple hens sharing the nest box. Often times when the hens are adjusting to make room for a new hen that wants to join the nest, eggs get stepped on. Even if the eggs are not broken, stepped on eggs become dirty eggs. If you frequently see two or more hens in the same box, you may want to consider adding more nest boxes.

Some poultry keepers say that curtains on the nest boxes curbs eat eating habits. I have curtains on my nest boxes to create a dark and relaxing environment and I have never had to deal with egg eating. I cannot be sure if there is a connection, but adding curtains certainly won't do any harm.

Curbing Egg Eating

As I stated earlier, most documentation (especially scientific research and studies) will state to cull the bird as soon as it's identified. If you have a large flock and need to keep egg production at its max, culling – or killing – may be your only option. If you have a smaller flock, there may be some options you can try. None of these methods has been scientifically supported, but they are deemed humane.

First, identify and remove the offending bird. This habit spreads quickly. Some poultry keepers have had a lot of luck by simply putting golf balls in the nest boxes and collecting the eggs frequently. The egg eater will assume the golf balls are eggs, peck for awhile and then give up trying to get the balls open. Other keepers have used a similar dummy egg method, filling plastic eggs with mustard, so when the chicken gets the eggs open, they are repelled by the taste. A similar method suggests filling plastic eggs with dish soap. This is cruel to the bird and may cause all sorts of digestive issues if any of the soap is consumed.

If you are handy, you can create roll out nest boxes. These boxes have a smooth liner placed in them. When the egg is laid, it rolls into an area that the chicken does not have access to.

A study conducted by the University of Florida (2011) showed a reduction in egg eating behavior when hens were allowed to drink fresh milk. The connection between the milk and the behavior is unclear, but may have something to do with calcium and protein deficiencies.

HUSBANDRY HINT

There is a product on the market called Peepers, plastic "clip-on" blinders. They were designed to prevent cannibalism, but they are effective in preventing an egg eater from pecking their eggs and can be decorated with drawn on eyes for an even sillier appearance.

INTERNAL PARASITES

Internal parasites are common among poultry. Parasites such as cecal worms are relatively harmless, while others, such as gapeworm, can greatly reduce egg production and become fatal if left untreated.

When it comes to internal parasites, biosecurity is the best preventative measure you can take. Constantly monitoring each member of your flock, with an action plan in place in the event a member shows signs of parasites, prevents further infection. As part of your action plan, you must decide whether you are most comfortable using commercial products or natural alternatives to treat your infected birds. This decision does not need to be an all-or-nothing approach. Like me, you may prefer to use natural options as a first line of treatments and have commercial products as a backup. There is speculation about the effectiveness of treating internal parasites with natural alternatives, making it a hot topic of debate.

Common Parasites

The most common internal parasite of chickens is coccidia, a protozoa frequently found where chickens are kept.[1] There are several recognized vaccines to prevent coccidia, including

CHICKEN HOT TOPICS

Advent®, Coccivac®, and Immucox®. Some poultry keepers refrain from giving coccidia vaccinations because the cost of vaccines can be high for the backyard poultry keeper and/or the possible side effect of vaccination may be the development of lesions and Coccidiosis, which would require antibiotics to treat secondary gut diseases. I have never vaccinated nor treated my flock for Coccidiosis. A clean environment can be the best way to prevent infection.

Some other common internal parasites of chickens include roundworms and thread worms. Folklore regarding the treatment of worms in livestock have floated around for generations. In recent years, researchers have begun studying the effectiveness of these treatments.

Garlic Treatment

A study by the University of Kentucky, College of Agriculture (2011) indicates garlic in combination with worm-repelling plants is effective in preventing intestinal parasites and worms. There worm-repelling plants include peppermint and wormwood. Allicin – a biologically active compound of garlic – appears to be effective against e. coli, staphylococcus aureus, clostridium perfringens and salmonella ssp.. Allicin kills rotovirus infection, which is responsible for many cases of diarrhea. The quantities required to treat poultry for these conditions has not been sufficiently studied. Be aware that large quantities of garlic in the diet may affect the flavor of eggs produced.

Pumpkin Seed Treatment

Pumpkin seeds have long been suggested for controlling tapeworms in chickens. In 2008, Delaware State University

Extension Program began a research study to investigate a number of natural plants (including pumpkin) in reducing fecal egg counts in goats. Pumpkin and many other vine crops are believed to contain cucurbitacin, which has been used to expel tapeworms and roundworms. The results of the first stage of the study showed pumpkin seeds were not effective in reducing fecal egg counts, however, they also observed that the goats were reluctant to consume the seeds. Further research is being conducted to examine the methods of administering the seeds as well as the long term results of the pumpkin seed study.

Diatomaceous Earth Treatment

Diatomaceous Earth (DE) has long been purported to control internal parasites in a variety of livestock. DE is silica powder created from fossilized aquatic organisms. In 2011, The Avian Research Centre of the University of British Columbia conducted a study regarding the effectiveness of DE as a treatment against internal parasites, its ability to increase feed efficiency, and boost egg production. The study was conducted with a 2% supplementation of DE being fed to two commercial laying breeds, Bovan Brown (BB) and Lowmann Brown (LB). The results of the study were quite perplexing. The LBs – known for being a parasite-resistant breed – showed little improvement in fecal egg counts when DE was added to their diet. The BBs, however, showed significantly lower fecal egg counts. Both breeds fed the supplement were considerably heavier, laid more eggs, and consumed more feed than the control birds. These results, as well as similar study results, leaves the effectiveness of DE as an internal antiparasitic still unclear.

Egg Withdrawal

A common concern regarding commercial products is the transference of the medication to the bird's meat and eggs. According to the Journal of Veterinary Pharmacology and Theraputics, *"poultry treated with pharmaceutical drugs can produce eggs contaminated with drug residues."*[2] The journal states that these residues could post a health risk to consumers. The time between medicating and safely consuming the eggs is called the Egg Withdrawal Period. The U.S. seems hesitant to approve drugs for laying hens, therefore egg withdrawal periods often are not listed on medications. This poses a challenge for poultry keepers who want to treat for parasites as well as other conditions that may require commercial medications. Because of this, poultry keepers are often reluctant to use commercial medications. The E.U. has approved a variety of medications for laying hens and continues to study egg withdrawal periods.

FUN FARMER FACT

From Dr. George Messenger, DMV, Dipl. ABVP-Avian:

"When we, as humans, refer to colds, we are generally talking about upper respiratory viral infections that humans get so commonly. There are quite a number of viruses that cause colds in people, with the resultant symptoms of runny nose, coughing, sneezing, nasal congestions, and others.

Chickens can, and often do, get upper respiratory infections. We call them colds, but just need to understand that they are not necessarily the same thing as a human cold. Chickens can become infected with a wider variety of pathogens, ranging from mycoplasma to bacteria to parasites. Some of these can become very serious and can often result in death of the animal and therefore it might not be proper to call them colds.

In summary, I would say that yes, chickens do get colds or cold-like illnesses, but most often they are not minor, self-limiting illnesses, and the cause of each respiratory illness in poultry should be investigated."

HUSBANDRY HINT

Collect pumpkins in your neighborhood after Halloween for winter-long treats. Store them on a pallet under a tarp for protection from the elements. The chickens will love your for it. Just be sure there is no candle wax residue.

CHICKEN HOT TOPICS

1   McMullin, Paul "A Poultry Guide to Poultry Health and Diseases" (2004)

2   Goetting, V. ; Lee, K.A. ; Tell, L.A. "Pharmacokinetics of Veterinary Drugs in Laying Hens and Residues in Eggs: A Review of the Literature" *Journal of Pharmacology and Therapeutics* v 34.6 (2011)

EXTERNAL PARASITES

Chickens get bugs. It's a fact of life. How a poultry keeper deals with these external parasites is a personal choice, based the on knowledge at hand. There are many commercial and natural options available, though the effectiveness of these options is sometimes in question.

Common Parasites

The two types of external parasites that affect chickens are Continuous and Temporary. As the names imply, continuous parasites spend their entire lives on the host and temporary parasites feed on the host, but live elsewhere. Temporary parasites often dwell in the chicken's bedding, roosts or coop. It is important to identify the external parasite so you know how to deal with it properly. If you have a temporary parasite and you only treat the bird and not the environment, you will not eradicate the problem. Common continuous parasites that affect chickens include the northern fowl mite, sticktight flea, scaley mite, and chicken lice. Common temporary parasites include fowl ticks, chicken mites, and bed bugs. Almost all external parasites are brought in from wild hosts like rodents and birds.

Illustration 3: Chickens dust bathe to rid themselves of external parasites.

Permethrin Treatment

Permethrin is a common commercial insecticide used for treating poultry parasites. Permethrin is created from a synthetic extract from chrysanthemums that has been chemically engineered to be more toxic with longer breakdown times.[1] Although permethrin is considered the most effective insecticide by poultry keepers when it comes to dealing with external parasites, that strength comes at a cost. Some forms of permethrin are classified by the U.S. Environmental Protection Agency (EPA) as Class C carcinogens. The greatest risk is through ingestion, so proper protection over the nose and mouth in the form of a mask can reduce risk.

Permethrin is available at many home and garden stores. A common brand is Sevin Dust®, which contains 0.25% permethrin as well as carbaryl, a moderately toxic insecticide. Some forms of the chemical are sold specifically for poultry, such as PoultryGuard™ and Prozap Insectrin Dust®. Both of these contain

0.25% permethrin, but lack the addition of carbaryl and therefore are safer choices. Both contain small amounts of formaldehyde.[2]

Pyrethrum Treatment

Pyrethrum is the natural extract from the flower head of chrysanthemums without chemical engineering. Pyrethrum is designated the strongest organically recognized insecticide and can be bought commercially under the name PyTGanic Pro®. Being an insecticide, pyrethrum can hardly be considered "safe," but the risks of using it are so low that "*a human can consume quantities of many grams without harm.*"[3]

Diatomaceous Earth Treatment

Diatomaceous Earth (DE) is thought to act as an insecticide by killing insects via desiccation, or drying out the insect. It is commonly used in homes to combat flea infestations. DE has proven itself an effective preventative measure against external parasites, but poultry keepers may require something stronger to combat a parasite infestation. To use DE as a preventative, sprinkle it inside the coop and nest boxes, as well as in dust bath areas.

Garlic Spray Treatment

A study conducted by University of Kentucky verified that the long-held belief regarding garlic as an external parasite remedy is valid.[4] They sprayed laying hens on the vent area with a solution of 10% garlic juice and 90% water weekly for three weeks. It decreased the northern fowl mites in the laying hens.

CHICKEN HOT TOPICS

---

1   "Chemical Fact Sheet: Permethrin" *Beyond Pesticides*

2   "Permethrin and Pyrethrin-Piperonly Butoxide" *Skin Therapy Letter* (2008)

3   Buss, Eileen A. ; Park Brown, Sydney G. "Natural Products for Managing Landscape and Garden Pests in Florida" *University of Florida IFAS Extension* ENY-350 (2013)

4   Jacob, Jacquie ; Pescatore, Tony "Natural Remedies for Poultry Diseases Common in "Natural" and "Organic" Flocks" *University of Kentucky Cooperative Extension* (2011)

WING CLIPPING

Wing clipping might seem intimidating. After all, you don't want to hurt your chicken. Never fear, wing clipping is as painless as clipping one's fingernails. Have you ever clipped your nails too short? There may be blood and a quick sharp pain, but there is no long-term damage. The same is true of a chicken's wing. If wing clipping is something you are considering, be sure to first educate yourself on how to properly perform the procedure.

Clipping vs. Pinioning

It is my belief that fear regarding the bird's psychological well-being stems from a confusion between wing clipping and pinioning. These are two completely different procedures. Wing clipping involves trimming the primary or (preferably) secondary feathers of one wing, throwing the bird off balance when it attempts flight. As mentioned before, wing clipping is not the least bit painful when executed properly and causes no long-term damage. Pinioning is a surgical procedure involving amputation of the part of the wing that grows the primary flight feathers. The Prevention of Animal Cruelty Act (1979) considers pinioning as *unreasonable, unnecessary* or *unjustifiable* as offenses like beating, kicking, maiming and torturing. Wing clipping and pinioning are certainly not the same thing.

Ethics of Reducing Flight Potential

The ethics of wing clipping has been greatly debated among bird lovers and the focus is solely focused on birds whose primary mode of mobility is flying. This is not the case with chickens. For most chickens, flight only takes place when they fly up to the security of their roost at night. One of the other reasons a chicken might take flight is to escape a predator. This brings us to the safety aspects of wing clipping.

If you choose to clip the wing of one or more of your birds, it is your responsibility to ensure their safety both inside and outside of the coop. Can your bird safely get on and off the roost? Can your chicken easily access the nest boxes? You, as their keeper, need to be aware of their daily patterns of where and when flight takes place so you can make sure accommodations are in place.

FUN FARMER FACT
Chickens have a variety of alarm calls. Each one tells the other members in the flock what danger is present and how to respond accordingly.

HUSBANDRY HINT
Do not clip the wing of a chicken that is prone to feather picking. Seeing the clipped feathers may further encourage the feather-picking behavior.

BREAKING A BROODY HEN

Broodiness is a hen's desire to sit on eggs and hatch them. When this behavior is taking place, the hen is referred to as a broody hen. Broody hens are great if you want to add to your flock naturally, but broody hens don't always have the best timing. Sometimes they would prefer to sit when you'd prefer them to lay eggs. When a hen begins "sitting," she ceases laying eggs.

Broody behavior is triggered by a buildup of the hormone prolactin. Production of prolactin seems to be connected to hot weather, dark corners in the coop, and seeing an accumulation of eggs. Broodiness can be avoided in some cases by keeping the coop well lit and the eggs collected frequently. If you have a hen or hens who frequently go broody, roll out nest boxes may be a good option for you. If eggs are out of sight, brooding may be out of mind.

If you notice your hen lingering on the nest after laying or spending a lot of time on the nest without producing eggs, she may be in the starting phases of brooding. If you do not want to hatch eggs, it is wise to remove her from the nest and relocate her. Be aware that she may act aggressively towards you when you try to handle her. She may need to be relocated several times before she gets the idea. Some hens – most notoriously Silkies – are so determined to hatch eggs that the simple act of removing her will have little to no effect.

If you allow a hen to hatch eggs (they can be chicken eggs, or they can be duck, goose or sometimes even quail eggs), she will hunker down until hatch day, only getting up once a day for a short time to eat and defecate. Some hens won't even get up once a day, choosing to stay on the nest for the entire incubation period. After incubation, the hen will raise her chicks and resume laying approximately five weeks after the chicks have hatched. If she is unable to hatch eggs, she has nothing to trigger laying and may "sit" for a longer time period than is healthy. A hen who broods gets pale in the comb, her feathers will get dull, and most importantly, she will lose weight from lack of eating. This is why poultry keepers choose to break their broodies. Starvation that lasts longer than normal incubation periods can be fatal.

Illustration 4: Hens make wonderful mothers and will raise almost any variety of poultry.

As cruel as "breaking a broody" may sound, it does not hurt the hen when it is performed responsibly. There is advice on the internet that encourages keepers to do a wide variety of cruel things to break broody hens. Some of these things include dunking the bird in cold water, and placing ice cubes under the hen while she sits on the nest. The flawed belief is that body temperature related to broodiness and by cooling the hen's body temperature, the broodiness will cease. Contrary to common opinion, the temperature of broody hens barely differs from that of laying hens.[1] Rapidly chilling the bird does nothing beyond stressing the bird's already stressed system.

A reliable method for stopping broody behavior is removing the hen from the flock and keeping her in a wire-bottomed cage or crate with no bedding. Again, poultry keepers commonly believe this method works because it cools the hen, but in fact is it because she cannot create a nest. It may be beneficial to add a supplemental light to increase light intensity while the hen is in the cage. Do not add additional hours of daylight, just intensity. A 25 watt incandescent bulb should be sufficient.

It can take anywhere from a few days to a few weeks for broodiness to subside. After the first three days, remove the hen and place her back in the flock. Watch for broody behavior, such as running for the nest boxes or aggression towards the rest of the flock. If she exhibits these behaviors, place her back in the cage for another three days. Once broodiness is broken, the hen will resume laying in one to two weeks.

A common question people have regarding breaking broodiness is if it hurts the hen emotionally. There is no research to indicate that hens suffer any emotional distress when not allowed to brood. Watch a broody after breaking and there will be no indication of heartache.

FUN FARMER FACT

Castrated male chickens can go broody when exposed to young chicks, indicating that broodiness is not restricted to females. Males will not, however, incubate eggs.

HUSBANDRY HINT

Do you want to allow your broody hen to raise chicks, but do not have access to fertile eggs? Many poultry keepers have had good results with giving their hens "foster chicks."

---

1   Hutt, F.B. "Poultry Genetics" *Salvat Editors* v. 1 (1958)

SUPPLEMENTAL LIGHTING

With shorter days comes less daylight in the winter, which affects the egg laying patterns of chickens. They will reduce their egg production or stop altogether during this period. Although eleven to twelve hours of light is sufficient to trigger laying, the amount is not adequate for continual high-production.

Some poultry keepers opt not to add supplemental lighting to their coops. They report feeling as though it is unnatural to "force" a bird to lay when their system deserves a break. Some of these naturalist keepers believe that supplemental lighting may decrease a hen's lifespan and cause the hen to stop laying prematurely.

Effect on the Hen

Dr. Mike Petrick, DMV/Msc states that there is no evidence to support the claims that supplemental lighting shortens a chicken's lifespan or harms it in any way, so long as they are provided adequate nutrition and allowed to molt every twelve to eighteen months. If the chicken molts – allowing it to replenish bone, muscle, and fat stores – it is equipped to healthily lay eggs. He adds that much like a human, female birds are born with all of their ova (eggs). Each hen has hundreds of thousands of ova and could not possibly lay long enough to run out. The bird will stop laying due to age, well before she runs out of ova.[1]

The first person known to use supplemental lighting was E.C. Waldorf, a medical doctor from Buffalo, New York, in 1889.[2] Despite results indicating that hens laid more eggs with supplemental lighting, it was twenty years before lighting chicken coops became routine.

I believe some of the negative feelings toward supplemental lighting stems from the fact that it's often abused in the factory egg industry. Egg factories are known to use systematic lighting patterns to force molt and increase growth rates. There is nothing natural about this practice.

Responsible Supplementation

The supplemental lighting I am endorsing is performed with the chicken's health and well-being in mind. That being said, when you do provide artificial light, it needs to be done responsibly. Do not stress your birds by adding four hours of light in the middle of December. Instead, increase the amount of light your bird receives by 15-30 minutes per week until you reach 14-16 hours per day (including natural daylight hours). Be advised that more than seventeen hours of daylight may actually decrease production and cause undue stress on your birds.

The best time to add supplemental lighting is in the morning hours. Birds do not object to sudden daylight, but a sudden decrease of light in the evening hours can disrupt roosting behavior. If you must add supplemental lighting in the evening hours, use a slow-paced dimmer to simulate sunset and trigger roosting.

Consistency is key when adding supplemental lighting. A drop in day length for even one day can be detrimental to egg production. A drop in day length for two days or more may cause your bird to go into a molt. Keep this in mind if you live in an area prone to frequent or long-term power outages.

Incandescent bulb have long been used for poultry lighting. One traditional 60 watt incandescent bulb is adequate for each 200 square feet of coop.[3] With energy efficiency in mind, many poultry keepers are switching to compact fluorescent bulbs. ENERGY STAR® qualified light bulbs supply bright, warm light that uses up to 75% less energy than incandescent bulbs and can last up to ten times longer. When selecting fluorescent lighting, choose warm wavelengths which emit red-orange rays. Cool wavelengths emit blue rays which are considerably less stimulating.

Safety Measures

Of course safety measures need to be address when installing lighting. Be sure that the electrical wiring meets all safety and local codes. Incandescent bulbs can get very hot and chicken feathers are highly flammable. Make sure to position the bulbs in a location the birds cannot access. As mentioned before, a sudden drop in daylight hours can negatively affect egg laying. It is recommended that you use an automatic timer and inspect it frequently to ensure it is operating properly. A final word of caution regarding the safety aspects of supplemental lighting, do not use bulbs labeled "Teflon-Coated," "Tefcoat," "Rough Surface," "Protective Coated," or "Safety Coated." The bulbs contain Teflon, which when heated, creates fumes that are toxic to birds and may result in illness or fatalities.[4]

Alternatives for Winter Production

If you desire ample egg supply through the winter, but are not comfortable with supplemental lighting, there are various breeds you can choose from that are recognized for their winter egg production. Heritage breeds such as Rhode Island Red, Delaware, Faverolle, New Hampshire Red, and Sussex produce well with limited lighting, as well as the hybrid Golden Comet.[5]

FUN FARMER FACT

Light is perceived by birds not only through their eyes, but through the skull. A blind bird will respond to daylight patterns in the same manner as a sighted bird.

HUSBANDRY HINT

Build your coop with windows positioned in a way that takes advantage of sunlight throughout the day.

---

1   Petrik, Dr. Mike "Supplemental Light in the Chicken Coop" *http://www.the-chicken-chick.com/2011/09/supplemental-light-in-coop-why-how/*

2   Hawes, Robert "Lighting For Small-Scale Flocks" *Cooperative Extension Publications* v. 2227 (2007)

3   Hawes, Robert "Lighting For Small-Scale Flocks" *Cooperative Extension Publications* v. 2227 (2007)

4   Lane, Jessica E. "Dirty Ovens & Greasy Stove Tops: A Warning About Teflon" *The 104 Homestead* http://104homestead.com/dirty-ovens-greasy-stovetops/

5   *Henderson's Chicken Breed Chart*

HEATING THE COOP

As winter approaches, poultry keepers begin to worry about the frigid temperatures to come. I get questions from readers as far south as Georgia about using supplemental heat during the winter. Being a Mainer, I find concerns of southern U.S. residents regarding their birds' comfort during the winter amusing, but I know that it is all relative. A cold day in Georgia is considered balmy here in the northeast. The average winter temperature in Maine is a chilly 30.9°F with stints of much lower temperatures. My Canadian and Alaskan readers might be thinking that's balmy to them. All things being equal, the poultry can be affected by extreme temperatures in a variety of ways, all negative. The chickens may become less active, eggs in the nest box and water in the coop may freeze, and combs and wattles have the risk of becoming frostbitten. It may be tempting to add a space heater or heat bulb.

Dr. Madeha Untoo, a member of Veterinary Sciences & Animal Husbandry at Shere Kashmir University, has researched the effects of extreme temperatures on poultry.[1] He has determined that while 65-75°F is optimal, poultry can effectively regulate their body in most extreme conditions. He states that temperatures exceeding 101°F are much more lethal than intense cold.

CHICKEN HOT TOPICS

Ventilation

The first and most important step in preventing winter discomfort for your chickens is to make sure there is proper ventilation. Although opening up spaces in your tightly built and well-insulated coop may seem counterproductive, air circulation is essential to keeping your birds healthy. Moisture trapped in the coop as well as ammonia build-up due to improper ventilation may cause respiratory issues. Fresh air circulating through the coop releases ammonia and dries out the air. Proper ventilation also greatly reduces the risk of frostbite on combs, wattles and toes, since frostbite occurs when moisture freezes on the skin.

Intentionally designed ventilation is not the same as drafts in your coop. All drafts – air that may leak in and cool the birds – needs to be addressed before cold temperatures arrive. Ventilation is most beneficial when it's located as high on the coop as possible so that lower level ambient temperatures stay the same. Drafts, on the other hand, cause fluctuations in temperature at the same height as the birds, making it very difficult for them to regulate their body temperatures.

Shock

Heat is not deemed essential for winter poultry care. In fact, an effort of compensating for extreme cold could be detrimental to your chicken's well-being. Intense fluctuations can cause shock to the birds' system. Some poultry keepers that are anti-heat cite power outages to be the greatest risk, but in fact, the greatest risk to chicken's health during extreme cold is the simple act of opening a coop door or allowing the chickens outdoor access that can cause fatalities. Going from a heated coop to a cold climate causes dangerous amounts of stress to the chicken's system.

Monitor your flock, bearing in mind that their capacity for cold is different from yours, and you will see from their behavior if there is a problem. Water heaters are available to keep water thawed, and you can insulate nest boxes and collect eggs frequently to avoid the eggs freezing. Fire risks and extreme temperature fluctuations make heating the coop an unsafe choice.

FUN FARMER FACT

In the days before electricity, farmers would warm rocks
by the fire and place them in stock tanks and water
dishes to keep them from freezing.

---

1   Untoo, Dr. Madeha "Management of Poultry in Extreme Weather" *Poultry Industry* (2012)

EUTHANASIA

Dispatching a chicken is rarely an easy thing to do emotionally. Many hobby poultry keepers never intend to dispatch a chicken. They plan to purchase chicks, grow them into healthy hens and then allow the hens to live out their days on the farm, go to another farm when they stop laying, or be sent to a meat processor for a more "hands-off" approach.

Whether or not you intend to cull (or dispatch) a bird, you need to be aware of how to do it. You need to plan for the unexpected. For example, what if a predator maims one of your birds without killing it and the bird is in pain? What if your bird has a health defect that causes suffering, such as cross-beak? Would you know how to end the chicken's life in a humane way? What is the most humane way?

Humane Euthanasia

The word euthanasia comes from the Greek words "eu," meaning "good," and "thanatos," meaning "death." A good death, in my book, is one conducted with the highest respect for the animal, that is swift, and causes little-to-no pain or distress. When it comes to poultry, there are easy ways to reduce fear and distress. Hold the bird gently in an upright position with the wings held to the sides to prevent flapping. The general public seems to be under the impression that hanging a chicken by the feet is acceptable. That position causes tonic immobility – the appearance of being temporarily paralyzed – which kicks in under extreme stress.

Cervical Dislocation

Cervical dislocation is the most humane way to euthanize a chicken, if the person performing the procedure is trained how to do it properly. Cervical dislocation separates the brain from the spinal cord and carotid arteries, causing rapid unconsciousness. This is not precisely "wringing the neck" or "breaking the neck," but rather a technique involving pulling to separate the spinal cord at the base of the brain within the upper third of the neck, providing a fast and painless death. There are many videos online that demonstrate a "broomstick method," but this technique is inhumane because it causes the bird extreme distress. If you are not strong enough to dislocate the neck and sever the spinal cord while cradling the bird, you may need to choose a different technique.

Carbon Monoxide

Carbon monoxide is the second most humane way to euthanize a chicken, but the cost and equipment involved may make it unsuitable for some poultry keepers. Carbon monoxide combines with the hemoglobin in the red blood cells, causing hypoxia.[1] Only pure carbon monoxide should be used. Car exhaust fumes are not acceptable. The car's catalytic converter removes most of the carbon monoxide from the exhaust and the fumes become so hot from the heat of the engine that the bird is put in severe pain and distress. Pure carbon monoxide gas should be piped into a sealed container by someone trained on the practice.

Gunshot

A gunshot directly to the head is deemed humane by The Center for Animal Welfare, University of California, and is a form of euthanasia that most poultry keepers can perform. The gun must be positioned to ensure that the brain is destroyed and it is recommended that the carotid arteries and jugular veins be severed immediately after the shot to make certain the bird is dead.[2]

Inhumane Methods

One popular form of culling – or killing – a chicken is in fact inhumane. This practice involves exsanguination and/or decapitation. Exsanguination is the bleeding out of a bird by slitting the neck and decapitation is the removal of the bird's head, traditionally performed with an axe and a chopping block. The exposed blood vessels resulting from exsanguination and decapitation may seal after being severed, delaying in unconsciousness and allowing for brain response for a brief period of time. This is especially true when using dulled equipment. Research conducted by the European Food Safety Authority (2004) shows that a massive injury such as the cutting of skin, muscle, trachea, esophogus, as well as arteries and veins, results in very significant pain and distress. The lack of behavioral response is most likely attributed to shock and blood loss, not "calmness".

Euthanasia is an uncomfortable topic for most and it is easy to brush the idea aside, but it is the poultry keeper's responsibility to be informed and have a plan should the need arise.

FUN FARMER FACT

Mike the Chicken earned his nickname "Miracle Mike" on September 10, 1945, when he was beheaded for a special dinner for his keeper's mother-in-law and Mike survived. He survived for not merely a day or two. Mike the Chicken lived for eighteen months post-decapitation, touring the country and earning $4500 a month in newspaper and magazine appearances. Mike's keeper fed him during those eighteen months using a grain and water mash in an eyedropper. Mike died of what may have been considered natural causes, had he not been previously beheaded, in March of 1947.

HUSBANDRY HINT

Make yourself familiar with local laws regarding livestock euthanasia before dispatching an animal. Some towns frown upon animal dispatching and in some cases, performing the act could constitute breaking the law.

---

1   Varon, Dr. Joseph; Marik, Dr. Paul E. "Carbon Monoxide Poisoning" *The Internet Journal of Emergency and Intensive Care Medicine* v. 1.2 (1997)

2   "AVMA Guidelines for the Euthanasia of Animals" *American Veterinary Medical Association* (2013)

DISPATCHING AGED BIRDS

Regardless of whether you decide to keep chickens as pets with benefits or to keep chickens as dual eggs and meat birds, layers end up in a special part of our hearts. With meat chickens, you know their fate right from the start. You can mentally prepare and because they grow out at a rapid rate, you don't have time to get to know them. Layers stay with us much longer. You know their personalities. You may have even given them names. That is what makes the decision of what to do when they cease laying very difficult.

If you have the space and finances to keep your aged chickens so they can live out their lives, that is wonderful. Unfortunately, it isn't a reality for most of us. The average chicken lives eight to fifteen years. Egg production does not begin until around six-months-old and tapers off quickly around two-years-old. Will your six-year-old hen still lay eggs? There is a chance, small as it may be. My three-year-old hen averages one blue egg every three weeks between the months of June and September. Either way, it is hard to finance a bird that no longer produces.

I will be completely honest. Being the highly practical person that I am, I had decided before getting my first chicken that I would keep them until they ceased laying and then we (meaning my husband) would dispatch them for meat. That was *before* getting my first chicken. That was before getting to know them. As I said, I have a three-year-old hen who barely lays eggs anymore. She is still with us.

Making a decision for dispatching or allowing a bird to live out its life is a hard one. Although treatment of aged birds is a hot topic of debate, it is a very personal choice and only you can decide what is best for you and your flock. It shouldn't be a matter of discourse between poultry keepers. I am quite sure that no people feel good about the decision to cull due to lack of productivity, but the decision is their own to make.

APPENDIX

Herbals for Chickens

Insect Repellents

Plant these herbs around your coop to repel insects, or dry leaves and hang inside the coop.

Catnip	Lavender	Rosemary
Fennel	Pennyroyal	Spearmint
Feverfew	Peppermint	Tansy

Rodent Repellents

Plant these herbs around your coop to keep rodents at bay, or dry leaves and keep in pouches around nest boxes and feed.

Lemon Balm	Mint

Relaxant

These are great herbs to dry and sprinkle in the nest boxes. Relaxed hens lay better and have less incidence of egg binding.

Dandelion	Lavender	Rose Hips
Dill	Lemon Balm	

Laying Stimulants

These herbs can be fed to chickens dry or fresh and may stimulate laying and increase egg production.

Borage	Garlic*	Nasturtium
Comfrey	Marigold	Parsley
Dandelion	Majoram	Sage
Fennel	Mint	Thyme

Internal Parasite Management

These herbs can be fed to chickens if you suspect internal parasites in your flock. They may help keep the parasite load manageable.

Garlic*	Sage	Tansy
Hyssop	Spearmint	Thyme
Nasturtium		

These herbal remedies have not been studied extensively enough to yield conclusive results of effectiveness.

* Garlic and other strong-flavored herbs and foods should be given in moderation because the flavor can effect the flavor of eggs produced.

The Big List of Poisonous Plants

Please bear in mind when you are reviewing this list, most chickens will avoid these plants. If you have them on your property, you need not worry too much. However, if you have a bird that dies mysteriously, you should check the plant for chew marks to determine if poisoning was the cause of death.

- Arum Lily
- Amaryllis
- Aralia
- Arrowhead Vine
- Autumn Crocus
- Avocado
- Azaela
- Baneberry
- Beans – Caster, Horse, Fava, Broad, Glory, Scarlet Runner, Mescal, Navy, Precatory
- Bishop's Weed
- Black Laurel
- Black Locust
- Bleeding Heart
- Bloodroot
- Bluebonnet
- Boxwood
- Braken Fern
- Buckthorn
- Burdock
- Buttercup
- Camel Bush
- Caladium
- Cana Lily
- Cardinal Flower
- Christmas Candle
- Clematis
- Clivia
- Cocklebur
- Coffee Bean
- Coral Plant
- Coriander
- Corncockle
- Coyotillo
- Cowslip
- Daffodil

CHICKEN HOT TOPICS

- Daphne
- Delphinium
- Davil's Ivy
- Elderberry
- Elephant Ear
- English Ivy
- Ergot
- Eucalyptus
- Euphorbia Cactus
- False Hellebore
- Fire Thorn
- Flamingo Flower
- Four o'Clock
- Foxglove
- Glottidium
- Golden Chain
- Heliotrope
- Henbane
- Holly
- Honeysuckle
- Horsetail
- Hyacinth
- Hydrangea
- Jack-in-the-Pulpit
- Jasmine
- Jimsonweed
- Juniper
- Larkspur
- Lily of the Valley
- Lobelia
- Locoweed
- Locusts
- Lords & Ladies
- Lupine
- Malanga
- Marijuana
- Mayapple
- Mexican Poppy
- Milkweed
- Mistletoe
- Mock Orange
- Monkshood
- Moonseed
- Morning Glory
- Mountain Laurel
- Myrtle
- Narcissus
- Nettles (raw)
- Peace Lily
- Periwinkle
- Philodendrons
- Pig Weed
- Poinciana
- Poinsettia
- Poison Ivy
- Pokeweed
- Potato Shoots
- Pothos
- Privet
- Pyracantha

- Rape
- Rattlebox
- Red Sage
- Rhododendrons
- Rhubarb Leaves
- Rosary Pea Seeds
- Skunk Cabbage
- Sorrel (Dock)
- Snow Drop

- Star of Bethlehem
- Sweet Pea
- Swiss Cheese Plant
- Tobacco
- Virginia Creeper
- Wattle
- Wisteria
- Yellow Jasmine
- Yews

Of course all of these plants are not fatally poisonous, and again, most chickens will avoid consuming them naturally. It is still wise to have a list on hand.

Medication Doses and Conversions

Make sure you are administering medications for the correct disease, parasite and/or condition. Confirm that the medication is appropriate for the age (chick, juvenile, adult) and purpose (eggs, meat, pet) of your bird.

Most medications offer doses in ounces (oz), milliliters (ml), cubic centimeters (cc), and teaspoons (tsp). The doses are generally based on the weight of the average hen at about 5 pounds (lbs).

Common Conversions

- 1 ounce = 30 cc = 6 teaspoons = 30 ml
- 1 cubic centimeter = 1 milliliter

Administration Methods

Orally: Oral medications are given directly by mouth. They can be administered using a syringe with the needle removed or with an eyedropper.

Intramuscular: Typically performed by a veterinarian, intramuscular medications are given with a needle and syringe into the muscle. The easiest place to give an intramuscular injection is in the thigh, making sure to go in 1/4".

CHICKEN HOT TOPICS

Subcutaneous: Subcutaneous medications are given with a needle and syringe just below the skin. The easiest place to give a subcutaneous injection is just in front of the bird's leg.

Topically: Topic medications are applied directly to the skin, often on an infected area, such as a scratch or wound.

In Water/In Feed: These medications are both the easiest and the hardest medications to give. If you are medicating a flock, dosages are hard to guess since feed and water consumption may be different for each member of the flock. For individual birds that have been separated in the flock, provide only the medicated feed/water and be sure to mix a new batch each day.

Recommended Reading

Biosecurity for Backyard Flocks, Donna K. Carver, DMV, PhD, ACPV

Chicken Coop Ventilation, www.backyardchickens.com/a/chicken-coop-ventilation-go-out-there-and-cut-more-holes-in-your-coop/

Composting and Using Backyard Poultry Waste in the Home Garden, Lance Ellis, Stephen Love, Amber Moore, and Mario E. de Haro-Marti

Factors Affecting Egg Production in Backyard Chicken Flocks, University of Florida IFAS Extension

Historical Perspectives in Poultry Nutrition and Feeding Management, Dr. Steve Leeson

Normal Behaviors of Chickens in Small or Backyard Flocks, Extension.org

Proper Light Management for Your Home Laying Flock, Sheila E. Scheideler, Extension Poultry Specialist

CHICKEN HOT TOPICS

Ten Secrets to a Successful Broody Hen and Chick Adoption, www.grit.com/animals/ten-secrets-to-a-successful-broody-hen-chick-adoption/

Using Sand in the Coop, http://104homestead.com/sand-in-coop/

Wing Clipping 101, http://104homestead.com/wing-clipping-101/

Credits

Illustration Credits

1. Rachel Arsenault of GrowAGoodLife.com
2. Brandon Sutter of LoneStarFarmstead.com
3. Oxfordian via Flickr
4. Steve Buissinne

Table Credits

1. Poultry Science Association, Inc.

Fowl Language

Welcome to Fowl Language, a glossary of terms
as they are used in this book.

amino acids: Any one of many acids that occur naturally in living
things and that include some which form proteins.

ammonia: A colorless gas that is a compound of nitrogen and
hydrogen and is the chief nitrogen-containing waste product of
many organisms.

arsenic: A white or transparent extremely poisonous oxide of
arsenic used especially in insecticides.

Aspergillosis: An infection or allergic response due to the
Aspergillus fungus.

biosecurity: A set of preventative measures to reduce the
transmission of infectious diseases.

broiler: A chicken raised for meat.

brood: To raise chicks or a group of young birds (such as chickens)
that were all born at the same time.

Bumble Foot: A bacterial infection and inflammatory reaction on
the feet of birds.

CHICKEN HOT TOPICS

Caged Layer Fatigue: A syndrome characterized by an inability to stand on their feet and fragile bones.

carbaryl: A chemical in the carbamate family used chiefly as an insecticide.

carcinogen: A substance that causes cancer.

cecal worms: A very common worm found in the lower digestive tract of chickens and causes little to no harm.

chicken lice: A form of lice specific to chickens that does not affect humans.

chicken mites: Pests that feed on the blood, feathers, skin, or scales of the bird.

clostridium perfringens: One of the most common causes of food poisoning.

coccidia: Small single-celled organisms that live in the intestinal tract.

Coccidiosis: A parasitic disease of the intestinal tract.

comb: The fleshy growth or crest on the head of a chicken.

Cross-Beak: A condition in which the top and bottom beaks do not align properly.

cucurbitacin: A class of biochemical compounds developed in order to defend themselves from herbivores.

cull: To reject from the group; kill.

desiccation: The process of extreme drying.

dispatch: Put to death.

e. coli: A bacterium in the shape of a short rod that sometimes causes an intestinal illness.

excretion: Uric acid and feces.

Fatty Liver Syndrome: A build-up of excess fat in the liver cells.

formaldehyde: A colorless gas that consists of carbon, hydrogen, and oxygen, has a sharp irritating odor, and when dissolved in water is used to disinfect or to prevent decay.

fowl ticks: A small soft-bodied tick that is found primarily on chickens and other domestic fowl.

gapeworm: A parasitic nematode worm that infects the trachea of chickens.

genetically modified organisms (GMOs): Plants or animals that have been genetically engineered with DNA from bacteria, viruses or other plants and animals.

Gut Disease: A disease which attacks the gut or stomach.

hemoglobin: The molecule in red blood cells that carries oxygen.

heritage breed: Traditional livestock breeds that were raised by our forefathers.

husbandry: A branch of agriculture concerned with raising domestic animals.

hybrid: A chicken of mixed breed or composition.

hypoxia: A condition in which the body or a region of the body is deprived of adequate oxygen supply.

incubate: To sit on eggs to hatch them by warmth.

CHICKEN HOT TOPICS

layer: A chicken raised for egg production.

lethargy: Abnormal drowsiness.

molt: To shed feathers with the castoff parts being replaced by a new growth.

morbidity: The quality of being unhealthful.

northern fowl mite: A blood-sucking mite that can be found on poultry both day and night.

omnivore: An animal that eats food of both plant and animal origin.

organic (bedding): Bedding consisting of natural matter – hay, straw, wood shavings and wood particles.

organic (feed): Rations made of 100% organic approved ingredients with no unnatural additives.

Peepers: Blinders fitted to the beaks of poultry to block their forward vision.

plate counts: A procedure that allows microbiologists to estimate the population density in a liquid sample.

respiratory distress: Difficulty in breathing characterized by wheezing, coughing and raspiness.

rice hulls: The hard protecting coverings of grains of rice.

Rickets: A disease of chickens in which the bones are soft and deformed due to an inability of the body to use calcium and phosphorus because of a lack of vitamin D.

roll out nest box: A nest box with an angled floor that rolls eggs away from the reach of the chicken.

roost (noun): A support on which birds sleep.

roost (verb): The act of settling for sleep.

Rotovirus: A contagious virus that can cause inflammation of the stomach and intestines.

roundworms: One of the most common parasite of the digestive tract.

salmonella ssp.: Any of a genus of rod-shaped bacteria that cause various illnesses (as food poisoning) in human beings and other warm-blooded animals.

scaly mites: Microscopic insects that live underneath the scales on a chicken's lower legs and feet.

soy: An annual Asian plant of the legume family widely grown for its edible seeds rich in oil and proteins, as food for livestock, and for soil improvement.

staphylococcus aureus: A bacterium that is frequently found in the chicken respiratory tract and on the skin.

stickstraight fleas: A flea that burrows into the skin (usually facial area) where they lay their eggs.

substrate: A term referring to the floor of a chicken house. Interchangeable with litter and bedding.

tapeworms: Long, flat worms that attach themselves to your chicken's intestines.

Teflon: Synthetic fluorine-containing resins used especially for nonstick coatings.

thread worms: A worm which infects the gut and lay eggs around the chicken's vent which causes itch.

tonic immobility: A behavior in which some animals become apparently temporarily paralyzed and unresponsive to external stimuli.

vent: The anus; the area of the chicken where excrement and eggs exit.

About the Author

Jessica studied zoology at Santa Fe Community College Teaching Zoo, but chose to trade the busy city life for the peace and quiet of rural living. She is now a full-time mom and backyard farmer. Jessica is dedicated to teaching individuals how to farm and homestead no matter where they live. This calling has led to a professional blogging career where she writes about simple living skills, real food recipes, gardening in small spaces, and so much more.

Come visit Jessica at http://104homestead.com for more information on chicken husbandry as well as some other great controversial topics.

www.ingramcontent.com/pod-product-compliance
Lightning Source LLC
Chambersburg PA
CBHW071247170526
45165CB00003B/1267